SCIENTIFIC AMERICAN EDUCATIONAL PUBLISHING

SCIENTIFIC AMERICAN INVESTIGATES THE HUMAN BODY

THE NERVOUS SYSTEM

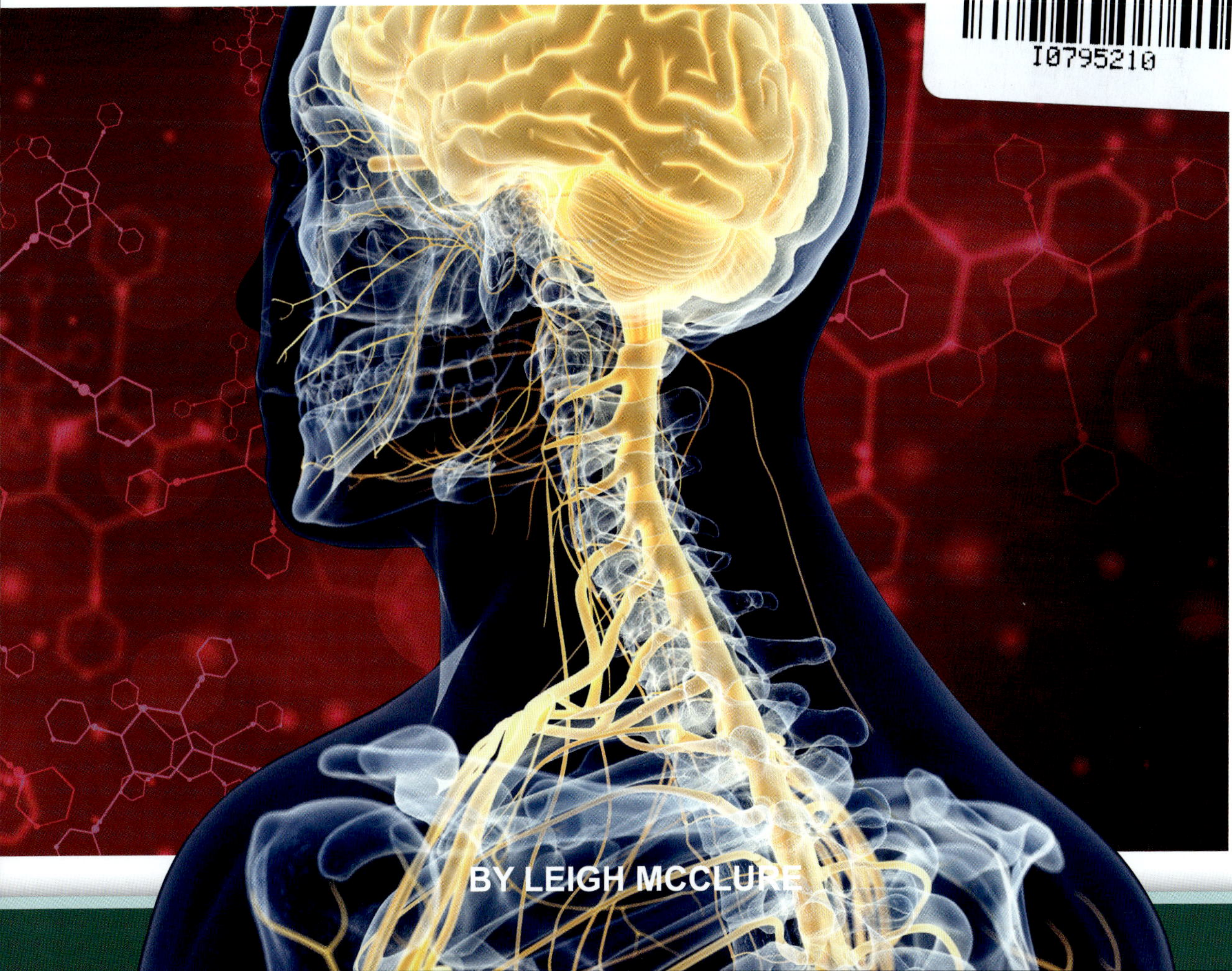

BY LEIGH MCCLURE

Published in 2025 by The Rosen Publishing Group
in association with Scientific American Educational Publishing
2544 Clinton Street, Buffalo NY 14224

Library of Congress Cataloging-in-Publication Data
Names: McClure, Leigh, author.
Title: The nervous system / Leigh McClure.
Description: Buffalo, NY : Scientific American Educational Publishing, [2025] | Series: Scientific American investigates the human body | Includes bibliographical references and index.
Identifiers: LCCN 2024011247 (print) | LCCN 2024011248 (ebook) | ISBN 9781725351455 (library binding) | ISBN 9781725351448 (paperback) | ISBN 9781725351462 (ebook)
Subjects: LCSH: Nervous system–Juvenile literature. | Brain–Juvenile literature.
Classification: LCC QP361.5 .M37 2025 (print) | LCC QP361.5 (ebook) | DDC 611/.8–dc23/eng/20240324
LC record available at https://lccn.loc.gov/2024011247
LC ebook record available at https://lccn.loc.gov/2024011248

Designer: Andrea Davison-Bartolotta
Editor: Caitie McAneney

Portions of this work were originally authored by Heather Moore Niver and published as *The Nervous System*. All new material in this edition authored by Leigh McClure.

Photo credits: Cover, p. 1 (main) SciePro/Shutterstock.com; cover, p. 1 (background) jijomathaidesigners/Shutterstock.com; p. 5 Roman Samborskyi/Shutterstock.com; p. 7 Brocreative/Shutterstock.com; pp. 9, 29 Monkey Business Images/Shutterstock.com; p. 11 VectorMine/Shutterstock.com; p. 13 Nandalal Sarkar/Shutterstock.com; p. 15 AYO Production/Shutterstock.com; p. 17 Iryna Inshyna/Shutterstock.com; p. 19 Dee-sign/Shutterstock.com; p. 21 Lorraine Swanson/Shutterstock.com; p. 23 BlurryMe/Shutterstock.com; p. 25 Motortion Films/Shutterstock.com; p. 27 John A. Anderson/Shutterstock.com; p. 28 GOLFX/Shutterstock.com.

Printed in the United States of America

CPSIA compliance information: Batch #CSSA25. For Further Information contact Rosen Publishing at 1-800-237-9932.

CONTENTS

Words in the glossary appear in **bold** type the first time they are used in the text.

YOUR OWN SUPERCOMPUTER

Do you know that you have a supercomputer in your own body? It's your brain! Your brain controls everything your body does. It allows you to use your five senses—sight, hearing, touch, smell, and taste. You can use those senses to understand what's happening around you and inside of you. That information determines how you will act or how your body will react.

The nervous system is made up of your brain, spinal cord, and nerves. Information speeds through these parts like it's on a highway. Your nerves send messages to the brain. Your brain decides the best course of action, and then it sends that command to the body. Everything you do each day, from walking to sleeping to eating, is thanks to your nervous system.

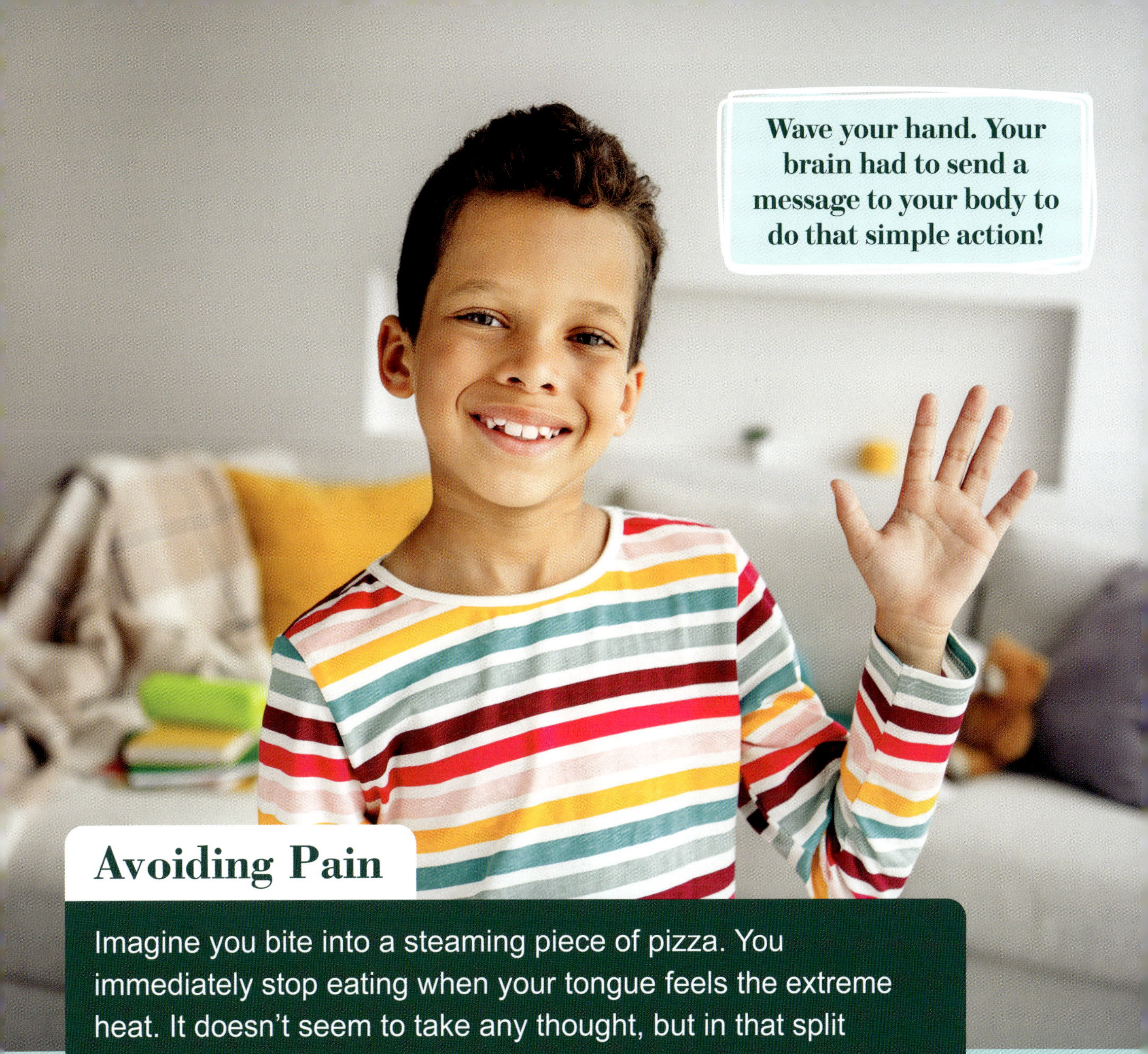

Wave your hand. Your brain had to send a message to your body to do that simple action!

Avoiding Pain

Imagine you bite into a steaming piece of pizza. You immediately stop eating when your tongue feels the extreme heat. It doesn't seem to take any thought, but in that split second, a lot of activity is going on in your body. Nerves in your tongue send a message about pain up to your brain. Your brain immediately sends a message to your hand to pull the pizza away. Burn avoided. Thanks, nervous system!

YOUR CENTRAL NERVOUS SYSTEM

The central nervous system includes your brain and spinal cord. Your brain is kind of like your body's central computer. It's the command center. It's in charge of every single action and reaction you have. And it acts in almost no time at all.

Your spinal cord is like a highway along which messages speed to and from the brain. The spinal cord is a long rope of nerve tissue that runs from the base of your brain down through the spine. On average, the spinal cord is around 18 inches (46 cm) long, and it's 0.5 inches (1.3 cm) at its thickest point. If you run your hand along your spine, you'll feel bumps. Those are **vertebrae**, which surround the spinal cord.

FUN FACT

ALONG THE SPINAL CORD, VARIOUS NERVES BRANCH OUT AND EXTEND THROUGH YOUR WHOLE BODY. THEY MAKE UP THE PERIPHERAL NERVOUS SYSTEM, WHICH TRANSMITS MESSAGES BETWEEN THE CENTRAL NERVOUS SYSTEM AND THE REST OF THE BODY.

How fast can the brain react to something? Neurons communicate with your brain at more than 150 miles (241 km) per hour.

Important Protection

The brain and spinal cord have a lot of important work to do. Of course, the body has some ways to protect them and make sure they stay healthy. Your skull protects your brain, and vertebrae are armor for the spinal cord. Layers of membranes, called meninges, and **cerebrospinal fluid** provide some padding. Cerebrospinal fluid also keeps the nerves healthy.

THE AUTONOMIC NERVOUS SYSTEM

You have to think about running a race or reading a book. But what about the things your body does without you thinking about them? Those things work because of the autonomic nervous system. This system regulates some body functions, like breathing and your heart beating, without your attention on them.

The autonomic nervous system is divided into two major parts: sympathetic and parasympathetic. Sympathetic nerves prepare your body to react to an emergency. They also kick in when your body's internal chemistry is out of balance. Parasympathetic nerves keep your body running when it's at rest. This system is often called the "rest and digest" system. When this system is turned on, you feel at ease.

When your parasympathetic nervous system is in charge, your body can easily process the food you eat.

Is That a Threat?

Sympathetic nerves help your body adjust to the world around you, such as when you sweat in hot weather. But when you face a threat or a difficult new challenge, the whole sympathetic nervous system kicks into high gear. Your lungs work harder and your heart pounds faster to send more oxygenated blood to your muscles. That way, you're ready to face the problem or quickly run for safety. This is called the "fight-or-flight response."

A PEEK AT THE PERIPHERAL

The autonomic nervous system is only one part of the peripheral nervous system. Two more groups of nerves are included—spinal and cranial. Spinal nerves start in the spinal cord. They contain bunches of sensory and motor fibers. Each connects the spinal cord to a certain part of the body. Motor fibers send information from the central nervous system to the muscles. Sensory fibers carry information from skin, muscles, joints, and internal organs to the spinal cord.

Cranial nerves join parts of the brain to sense organs in your head. They also connect to muscles, internal organs, and **glands** in the head and upper body. Cranial nerves are hard at work when you smile, sing, or eat an apple.

FUN FACT

HUMANS HAVE 31 PAIRS OF SPINAL NERVES.

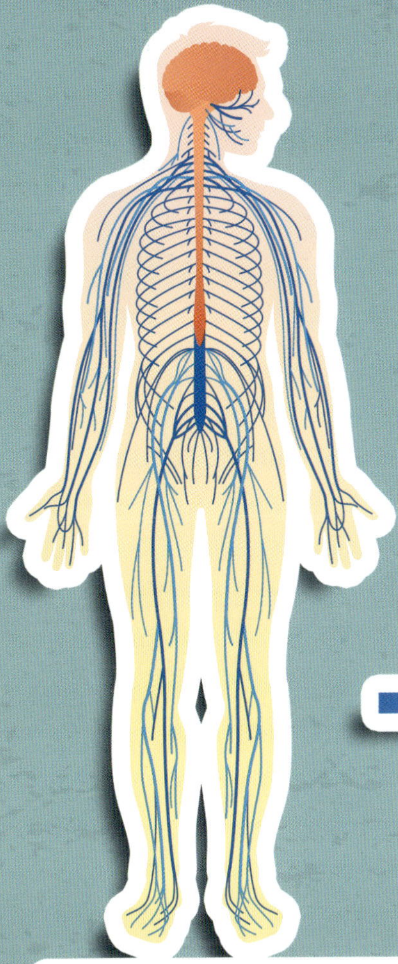

This diagram shows how the peripheral nervous system sends messages from the brain, down the spinal cord, and to nerves all over the body.

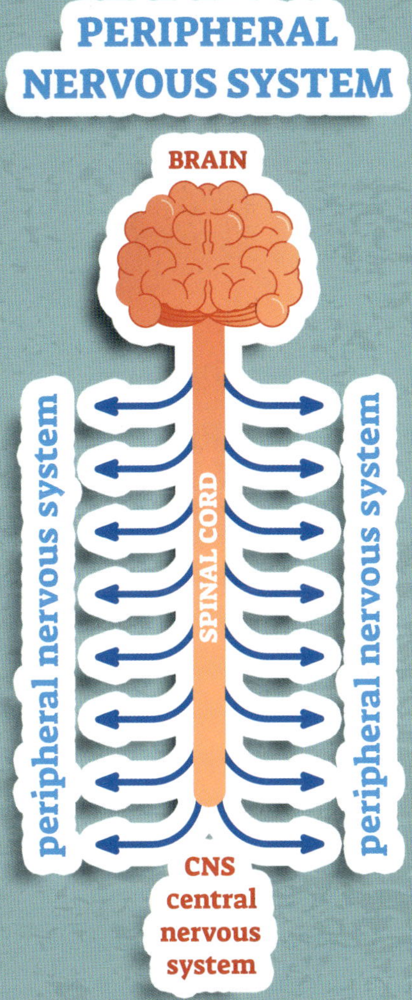

Time to Grow!

At the base of your brain is a tiny gland called the pituitary (or hypophysis). It makes and releases special **hormones** to make your body grow. It's also responsible for releasing hormones that prompt **puberty**. Your hypothalamus is near the pituitary gland in your brain. It produces hormones that control many things, such as body temperature, mood, thirst, hunger, and even the release of hormones from other glands.

PARTS OF THE BRAIN

When you imagine a brain, you may imagine one big wrinkly lump. But it's actually made up of three major sections: the forebrain, midbrain, and hindbrain. The forebrain is the biggest and most complicated part. It's mainly made up of the cerebrum, which contains the information that gives each of us a personality—emotions, intelligence, and memory. It also allows us to feel and move around. Five areas of the cerebrum, called lobes, are responsible for different kinds of information storage and activities.

The cerebrum is divided into two halves, or hemispheres. The right half controls the left side of the body, and the left half is in charge of the right side of the body. The two sides communicate using a band between them, called the corpus callosum.

FUN FACT

THE FIVE LOBES OF THE CEREBRUM ARE CALLED FRONTAL, PARIETAL, TEMPORAL, INSULAR, AND OCCIPITAL.

Check out the lobes of the cerebrum with this diagram. The insular lobe is located deep inside the brain, covered by the other lobes.

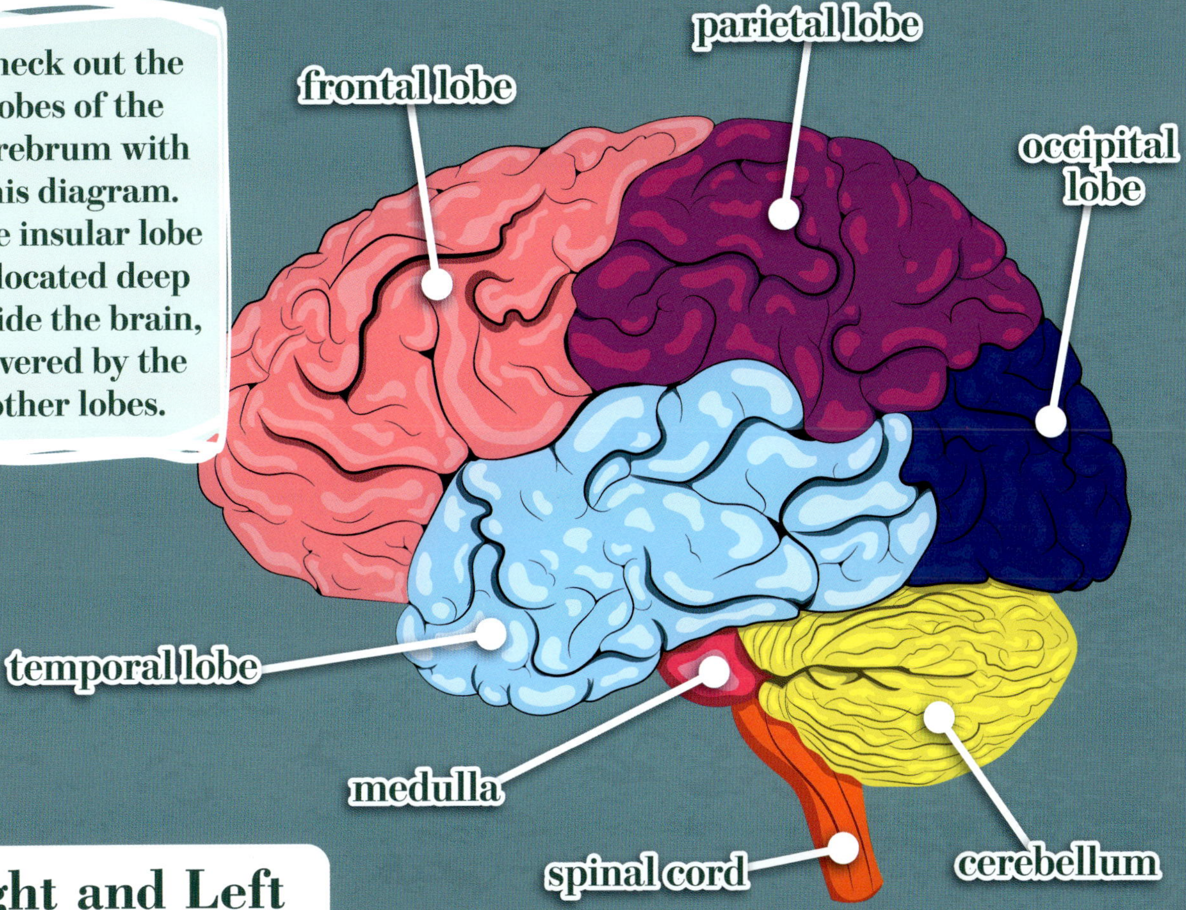

Right and Left

The cerebrum's two hemispheres look a lot alike, but they do more than control different sides of the body. The right side of your brain is sensitive and creative. The left side is more logical and objective. So when you're grooving to your favorite music, the right side of your brain is in action. When you do your math homework, the left side is busy. Creative people are often considered more "right brained," while scientists are often considered "left brained." In the medical world, this is mostly myth, however.

The cortex is the outermost layer of the cerebrum. This part is important because it collects information from your five senses and then sends that information out to other parts of the nervous system.

The thalamus, hypothalamus, and pituitary gland make up the inner part of the forebrain. The thalamus receives information from our sensory organs and passes it along to other parts of the brain.

The midbrain is below the middle of the forebrain. Its chief job is to manage the messages sent between the brain and spinal cord. Finally, the hindbrain sits below the back of the cerebrum. Part of it helps you balance and move. The cerebellum is a part of the hindbrain. Its name comes from the Latin words for "little brain" because it looks like a small cerebrum.

FUN FACT

THE HYPOTHALAMUS REGULATES MANY BODY PROCESSES THAT OCCUR AUTOMATICALLY. THE HYPOTHALAMUS ALSO CONTROLS THE PITUITARY GLAND.

You don't have to think about your heart beating for it to beat. The brainstem helps with this.

The Brain's Secretary

Together with the midbrain, parts of the hindbrain (called the pons and medulla) make up the brainstem. The brainstem's job is to receive, send out, and organize all the brain's messages. It's a little like the brain's secretary. It's also the part of the brain that helps with some of those actions that happen without any thought, like swallowing, blinking, breathing, and heartbeat.

LITTLE NEURONS, BIG JOBS

The last part of the central nervous system is the neurons. Neurons are shaped like flat, stretched stars. Many are less than 0.04 inches (1 mm) long. These nerve cells are the real stars! When you see, smell, taste, and feel ice cream, your sensory neurons send that information up to your brain right away. Motor neurons from the brain carry information telling your body to chew, swallow, and maybe even reach for more ice cream.

Neurons have four parts. The cell body is much like the cells in other parts of your body. The axon is a long, thread-like part that carries electrical impulses. The branched ends of axons, called axon terminals, send information to other neurons. Dendrites reach out toward other neurons to grab information. They look like branches.

Your skin has more than 4 million sensory receptors, which take in information about touch, temperature, and pain. Many of them are in your fingers and lips.

One Outstanding Organ

An adult brain only weighs about 3 pounds (1.4 kg), but it controls every single thing you do. With plenty of nooks and crannies, it has the extra surface area to save all the body's important information. Your brain contains more than 100 billion neurons, and they're so small you need a microscope to see them. Each one has a tiny axon and tiny dendrites reaching out to other neurons.

Synapses are the places where neurons meet and communicate. They do this all the time! One neuron may form synapses with 5,000 to 200,000 other neurons.

Neurons have two kinds of synapses. Electrical synapses, or gap junctions, are physical connections between the dendrites of different neurons. They allow electrical impulses to pass freely from one dendrite to another. Chemical synapses, or synaptic clefts, are spaces between the axon terminal of one neuron and the dendrite or cell body of another. When an electric signal reaches the axon terminal, the terminal releases a special chemical that crosses the synaptic cleft. The dendrite receives the chemical message and turns it back into an electrical impulse. This must happen again and again until the message gets to where it needs to go.

FUN FACT

THERE ARE LITTLE CLUSTERS OF CELLS ON EITHER SIDE OF YOUR BRAIN CALLED AMYGDALA (UH-MIHG-DUH-LUH). SCIENTISTS THINK THEY CONTROL YOUR EMOTIONS.

Neuron Network

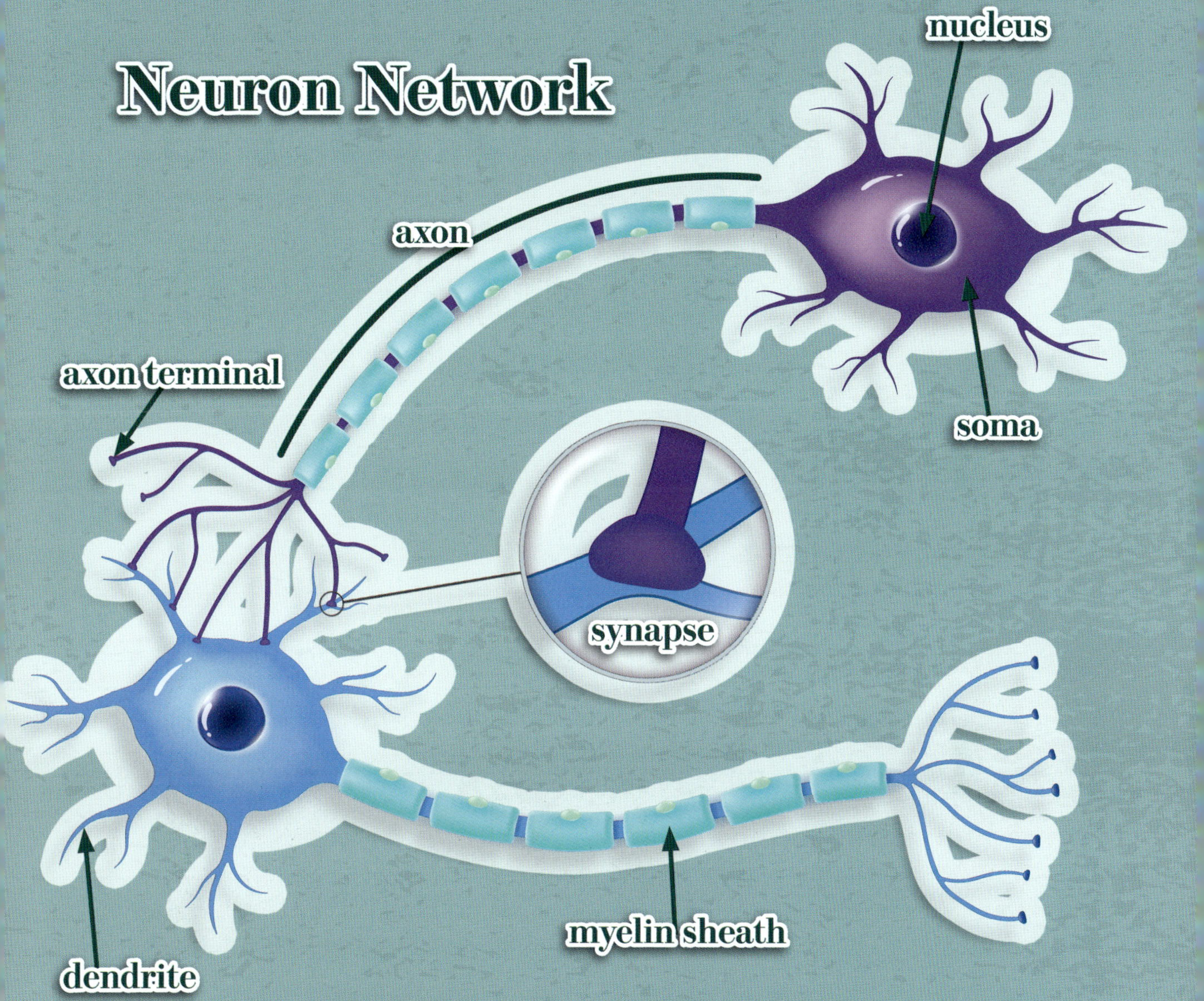

Explore the path that information takes to get from one neuron to the next.

LEARNING AND CHANGING

You're born with most of the neurons you'll ever have. However, they don't have connecting pathways at first. Your brain makes them as you learn. You have billions of neurons that help you learn. As you learn more, the messages move between neurons over and over again. Once you do something enough, your brain notices patterns and creates pathways between the neurons to make things easier.

Think about how hard it can be to learn how to do something new, like a dance routine. At first, it can be really hard to remember how to move your arms and legs to the beat. But as you practice, your neurons send the same messages again and again. Eventually, your brain creates a "dance routine" pathway to help you.

FUN FACT

LEARNING THE MOTIONS REQUIRED TO DO SOMETHING IS OFTEN CALLED MUSCLE MEMORY. THE MORE YOU DO SOMETHING, THE BETTER YOUR BRAIN REMEMBERS HOW TO MOVE YOUR MUSCLES DURING THAT ACTIVITY.

The first time you skate, it may be hard to figure out how to move your feet and keep your balance. Over time, your body creates neural pathways to make it easy!

Finding Balance

The cerebellum is at the back of your brain. It's only a fraction of the size of the cerebrum, but it has a really important job to do. Balance, movement, and how your muscles work together—or coordination—are all controlled by your cerebellum. When you stand upright and walk around without toppling over, thank your cerebellum.

PROBLEMS WITH THE NERVOUS SYSTEM

Doctors who help with nervous system disorders are called neurologists. Problems with the nervous system can show up as little changes in personality, more serious **symptoms** like blindness, or even death. Some disorders are **hereditary**, while others result from an injury or **infection**.

A traumatic brain injury (TBI) can cause memory and mood problems. An injury to the spinal cord can cause paralysis. Infections can cause meningitis, which is an inflammation of the protective **membranes** of the brain and spinal cord. Polio attacks motor neurons, and many children became paralyzed from it until a **vaccine** was created. Ticks spread germs that cause neurological diseases. Lyme disease, for example, is spread by deer ticks.

FUN FACT

SHINGLES, OR HERPES ZOSTER, CAN OCCUR IN SOMEONE WHO HAS BEEN INFECTED WITH THE CHICKEN POX VIRUS. IT ATTACKS THE NERVE ENDINGS AND CAUSES A RASH ON THE SKIN.

Injuries to the frontal lobe of the brain can cause changes in personality.

Dealing with MS

Multiple sclerosis (MS) cripples the central nervous system. It's called a demyelinating disease because it breaks down the **myelin sheath**. If the myelin sheath is damaged, nerve fibers can't work correctly. The **immune system** is often affected. Certain white blood cells attack the myelin as if it were an enemy. MS is the most common demyelinating disease, and its cause is unknown. Scientists are searching for a cure. In the meantime, people with MS often take medicine and do some physical therapy.

A HEALTHY BRAIN

It's important to keep your brain and the rest of your nervous system healthy. This system affects everything you do! You can exercise your brain and give it the right fuel.

Exercise that's good for your body is also good for your nervous system. Go outside and play! Try new things, like hiking a new path or trying a new skateboarding trick. Always be sure to wear a helmet when you ride a bicycle or play any sport where you might bump your head. Avoid drinking alcohol, taking drugs, and smoking. Also, be sure to use your brain a lot by doing puzzles, reading books, and even painting or drawing. You should always eat a healthy diet with lots of food containing potassium and calcium. These minerals help your nervous system function properly.

Strategy games like chess are great for exercising your brain.

Puzzle Time!

If you really want to give your brain a tough workout to keep it in shape, try Sudoku. It's a number puzzle that uses three rows of three boxes. Some of the numbers are there, and you have to figure out the missing numbers. You could also try getting all the colored squares on one side of a Rubik's cube or even on all the sides! Other ideas are crossword puzzles and chess.

A LOOK AT ANIMALS

Most animals on Earth have some kind of nervous system. Animals with backbones have similar nervous systems to ours. These animals are called vertebrates, and they include fish, birds, reptiles, amphibians, and other mammals. Generally, they have a brain and spinal cord that do a lot of the "heavy lifting." They also have a peripheral nervous system that carries messages between the brain and spinal cord and the rest of the body.

Invertebrates, or animals that don't have a spinal cord, have different kinds of nervous systems. Many invertebrates have a nerve net, or a system of nerve cells and fibers that are distributed across the body like a net. The movement of electrical impulses is much slower in animals with a nerve net.

FUN FACT

JELLYFISH HAVE A NERVE NET BUT NO BRAIN.

The fish in a coral reef have a full nervous system like we do. But the sea sponge doesn't have a clear-cut nervous system at all!

Brainless Beings

Coral looks like it's not alive, but it's actually a sea creature. Though it's hard to believe, the coral animals you see in the sea have a nervous system. They don't have a brain like we do, but they do have a basic nerve net. The net reaches from the coral's mouth to its tentacles. Some coral even have simple senses, like taste and smell. These senses help the coral know if there is prey nearby.

LOVE YOUR NOGGIN

French philosopher René Descartes once said, "I think, therefore I am." That means that we know we're alive because we have the ability to think. And we have the ability to think because we have a nervous system.

The brain, spinal cord, and nerves allow you to do everything from breathing to reading this book. That means you need to keep your nervous system safe. Protect yourself from risky falls and hits while playing. Wear a helmet to protect your head while you're riding your bike or skateboarding. Eat foods with many nutrients and drink plenty of water. Your noggin contains your whole world. It allows you to sense your surroundings and act upon them. You only get one, so keep it healthy and safe!

Falling on your head could cause a traumatic brain injury. Always wear a helmet.

GLOSSARY

cerebrospinal fluid: A clear liquid that comes from the blood and helps cushion the brain and spinal cord.

gland: Organs in the body that make and release hormones.

hereditary: Passed from parent to child.

hormone: A chemical made in the body that tells another part of the body how to function.

immune system: The parts of the body that fight germs and keep it healthy.

infection: The spread of germs inside the body, causing illness.

membrane: A thin tissue in the body.

myelin sheath: A protective layer around the axon of a nerve cell.

puberty: The stage or age at which a person experiences the maturation of the reproductive system.

symptom: Something that suggests a disease or physical problem.

vaccine: A shot that keeps a person from getting a certain sickness.

vertebrae: The ring-shaped bones that make up the spine.

FOR MORE INFORMATION

Books

Editors at Scientific American, Inc. *It's Mental: 10 Fun Memory and Psychology Experiments.* Buffalo, NY: Scientific American Educational Publishing, an imprint of Rosen Publishing, 2023.

Kehoe, Rachel. *Detecting Brain Disorders.* Lake Elmo, MN: Focus Readers, 2024.

Mason, Jenny. *Nervous System.* New York, NY: Children's Press, 2024.

Websites

Brain and Nervous System
www.childrensmn.org/educationmaterials/parents/article/7672/brain-and-nervous-system/
Dive deeper into the nervous system with this informative site.

Nervous System
kids.britannica.com/students/article/nervous-system/276055
Discover useful diagrams about the nervous system.

Your Amazing Brain
kids.nationalgeographic.com/science/article/your-amazing-brain
Explore amazing facts about your body's command center—the brain.

Publisher's note to educators and parents: Our editors have carefully reviewed these websites to ensure that they are suitable for students. Many websites change frequently, however, and we cannot guarantee that a site's future contents will continue to meet our high standards of quality and educational value. Be advised that students should be closely supervised whenever they access the internet.

INDEX